ANIMAL RECORD BREAKERS

WHALE SHARK

THE LARGEST FISH

THERESE M. SHEA

PowerKiDS press
New York

Published in 2020 by The Rosen Publishing Group, Inc.
29 East 21st Street, New York, NY 10010

First Edition

Editor: Elizabeth Krajnik
Book Design: Reann Nye

Photo Credits: Series art (frame) HiSunnySky/Shutterstock.com; series art (banner) Roberto Castillo/Shutterstock.com; series art (background) Wilqkuku/Shutterstock.com; cover, p. 13 Krzysztof Odziomek/Shutterstock.com; p. 5 22August/Shutterstock.com; p. 7 Onusa Putapitak/Shutterstock.com; p. 9 wildestanimal/Shutterstock.com; pp. 11 (whale shark), 21, 22 Rich Carey/Shutterstock.com; p. 11 (basking shark) Martin Prochazkacz/Shutterstock.com; p. 11 (megamouth shark) The Asahi Shimbun/Getty Images; p. 15 VisionDive/Shutterstock.com; p. 17 magnusdeepbelow/Shutterstock.com; p. 19 weera bunnak/Shutterstock.com.

Library of Congress Cataloging-in-Publication Data

Names: Shea, Therese, author.
Title: Whale shark : the largest fish / Therese M. Shea.
Description: New York : PowerKids Press, [2020] | Series: Animal record breakers | Includes index. | Summary: "Can you imagine swimming with the largest fish in the ocean? Some people have had this amazing experience! These fantastic fish are called whale sharks, and they can be more than 45 feet (13.7 m) long. Whale sharks may sound fierce, but they're only hungry for fish-not people. Readers will love reading about these slow-swimming gentle giants. Captivating photographs reveal just how massive whale sharks are, while fact boxes provide readers with additional interesting information"– Provided by publisher.
Identifiers: LCCN 2019030291 | ISBN 9781725308824 (paperback) | ISBN 9781725308848 (library binding) | ISBN 9781725308831 | ISBN 9781725308855 (ebook)
Subjects: LCSH: Whale shark–Juvenile literature.
Classification: LCC QL638.95.R4 S54 2020 | DDC 597.3/3–dc23
LC record available at https://lccn.loc.gov/2019030291

Manufactured in the United States of America

CPSIA Compliance Information: Batch #CWPK20. For Further Information contact Rosen Publishing, New York, New York at 1-800-237-9932.

CONTENTS

COLOSSAL CREATURE

Imagine swimming in the ocean. You see a shadow beneath you. It gets bigger as it gets closer. Now it's huge! Is it a whale? No, it's a whale shark!

A whale shark is a shark, which is a fish. The whale shark is the largest fish in the world. It's probably the largest fish ever to have lived on Earth. Most whale sharks grow between 18 and 33 feet (5.5 and 10.1 m) long. Some grow as long as 60 feet (18.3 m)!

ANIMAL ACTION

How big is a whale shark? Picture your school bus. This fish can be even longer!

The whale shark is the only member of its animal family!

A BIG, BIG BODY

The whale shark's body isn't what we picture a shark's body to look like. It has a wide, flat head with a huge mouth. Five **gill** openings can be found behind each side of its head. Hard **ridges** run to its tail. It has fins on the top of its body as well as on its tail.

Whale sharks have stripes and spots all over their bodies. A whale shark's pattern of markings is like a fingerprint. No two are exactly alike.

A whale shark's mouth is at the front of its body. Most other sharks have mouths on the underside of their bodies.

FINDING THIS FISH

Whale sharks can be found in warm waters around the world. Near the Americas, this includes the Atlantic Ocean, off the coast of New York and south to Brazil, and in the Pacific Ocean, from southern California to northern Chile. Whale sharks are also found in other parts of the Atlantic, Pacific, and Indian Oceans.

Whale sharks migrate, or travel, to certain places each year. They seem to have favorite spots to find food. Some swim thousands of miles!

ANIMAL ACTION

Sometimes as many as 800 whale sharks gather to eat near Mexico's Yucatan Peninsula!

Whale sharks usually live alone, but they can also be found in large groups.

PLENTY OF PLANKTON

Whale sharks have 300 rows of little pointy teeth! That sounds scary, but this fish doesn't tear its food apart like other sharks.

Here's how it eats: it opens its mouth and lets water—and anything in the water—pour in. It has body parts like a **sieve**, catching tiny animals and allowing water to escape out into the ocean. Then, the whale shark swallows whatever's left. The sharks mostly eat **plankton** but also other tiny fish or shellfish that get caught in their "nets."

Because of the way it eats, the whale shark is called a filter-feeding shark. A filter is a tool used to collect bits of matter from a liquid or gas.

SIZING UP FILTER-FEEDING SHARKS

WHALE SHARK
60 feet
(18.3 m)

BASKING SHARK
46 feet
(14 m)

MEGAMOUTH SHARK
15 feet
(4.6 m)

BIG FAMILIES

There's still much scientists don't know about how whale sharks are born. However, they do know whale shark mothers carry their eggs in their bodies until their young are ready to hatch, or break out. A female whale shark can carry more than a dozen eggs. They probably don't all hatch at once.

It's likely that shark pups, or babies, come out of the mother whale shark fully formed. The pups may be about 1.8 feet (0.55 m) long when they're born.

ANIMAL ACTION

A female whale shark caught near Taiwan had nearly 300 babies inside her! That's two times as many as other kinds of sharks.

Scientists think that a whale shark can live to be 60 to 100 years old.

MYSTERIOUS FISH

Whale sharks remain one of the most mysterious fish in the ocean. For example, in the places where many whale sharks come together to feed, most are young males. Scientists wonder where the older male and female whale sharks are.

No one has ever seen a female whale shark give birth. We don't know if mother whale sharks travel to a certain place to do this. Scientists hope to find answers to these questions by using **technology** to follow the sharks.

ANIMAL ACTION

Female whale sharks about to give birth are often seen near the Galápagos Islands off the west coast of South America.

Scientists can study a whale shark's food by looking into its big mouth.

DON'T BE AFRAID!

Whale sharks are only interested in eating plankton and other tiny ocean animals, so people don't need to be afraid of them. Since whale sharks look for food near the water's surface, people sometimes find themselves swimming around these gentle giants. Divers have even "ridden" these giant swimmers!

Whale sharks have bumped boats, either by accident or because they're interested in the bait fishermen are using. Unfortunately, sometimes whale sharks are hit by boats or boat **propellers**, which can harm them.

Whale sharks have approached swimmers, but scientists think it's simply because they're curious!

ENDANGERED!

No one is sure how many whale sharks are in the oceans. The guess is in the tens of thousands. While that may sound like a lot, the number worldwide is thought to have dropped by half in just 75 years.

In 2016, the whale shark was placed on the endangered species list. An endangered species is a kind of animal in danger of dying out. Laws protect the animals on this list. Still, whale sharks are hunted for meat in places like Asia.

ANIMAL ACTION

Whale sharks are used to make **medicine** in Asia.

Laws make it illegal to catch whale sharks. Some fishermen still do it because each fish provides a lot of meat.

MORE DANGERS

Besides hunting and boat accidents, whale sharks face other dangers. Pollution in water can kill their food source, affecting their health. Whale sharks also eat pollution, like bits of plastic or **chemicals**, when they eat what's in the water.

Scientists think sound pollution can harm whale sharks as well. The fish listen for **vibrations** in the water to know where food is. Loud noises from ships and underwater machines can make it harder to know where food is.

ANIMAL ACTION

Some businesses take people to feed whale sharks. Feeding them changes their migration patterns, which is also unhealthy for whale sharks.

Whale sharks' mouths are a bit like a **vacuum**. They suck in water, food, and whatever else is mixed in.

SEEING IS BELIEVING

It's important to keep whale sharks safe because they're beautiful creatures! To truly understand their size, you can visit them in some aquariums. In a few, you can even swim with these fantastic fish.

Scientists say that whale sharks are useful, too. They're a sign of a healthy ocean. That's because these sharks swim in areas with plankton, and plankton are found in healthy waters. We need healthy oceans for good food, clean air—and for swimming and having fun!

chemical: Matter that can be mixed with other matter to cause changes.

gill: The body part that animals such as fish use to breathe in water.

medicine: Matter used to lessen pain or to treat an illness.

plankton: A tiny plant or animal that floats in the ocean.

propeller: A machine part with two or more blades that turn quickly and cause a ship or plane to move.

ridge: A raised part or area.

sieve: A tool that has many small holes used to separate solids from liquids.

technology: A method that uses science to solve problems and the tools used to solve those problems.

vacuum: A machine that sucks up dirt.

vibration: Small, quick movements.

Due to the changing nature of Internet links, PowerKids Press has developed an online list of websites related to the subject of this book. This site is updated regularly. Please use this link to access the list: www.powerkidslinks.com/arb/shark